GRAf IT ART

Book 1: The Basic Skyline

Tan Huynh

Published by:
OClassrooms Education Services, LLC
8064 Allison Avenue #1184
La Mesa, CA 91944
www.oclassrooms.com

ISBN-978-0-9975696-0-5

Cover:
Designed by Tan Huynh
Gra*f* It Art artwork used with permissions by Jean Olmstead
Graphed using Desmos.com

Printed in the United States of America.

Contents

Acknowledgments

My success as a teacher has been the result of countless influences from all the teachers in my life, starting with my parents and family. There's no such thing as a "self-made person," only a "teacher-made person." My first acknowledgment goes to all the teachers in my life who have made a difference!

The birth of this project started with an art teacher named Andrea Barret who was willing to take a chance with my idea. She agreed to teach the art aspect of the project while I handled the mathematics side. The project was a tremendous success leading to an exhibition of the mathematics and artwork installed at a local art gallery for public viewing. Thank you, Andrea!

For two years, the Friends of East County were able to contribute generously to the Gra$f$ It Art project. I won two grants totaling \$2,000 for various interdisciplinary projects. The Gra$f$ It Art project costs roughly \$800 if an additional art component is added to it. This book focuses on a revised version that can be done at no cost. Thank you, Friends of East County!

Projects need a place to be displayed. It is this public viewing that raises the standards and expectations for students. When they know that it will be up for public viewing, they will do a better job. Sophie's Art Gallery was generous in time, space, and labor as they installed all of my students' projects for viewing. Thank you, Sophie's Art Gallery!

I'm blessed to be living in a time of great technology. However, when it comes to creating projects requiring technology, such as the technology to graph hundreds of lines, it can be a nightmare because different students have different technology platforms at home. I found Desmos.com, and it has been a life saver. It is cross-platform (Windows, Mac, Linux, and even all smartphones) and free for educational use. Thank you, Desmos.com!

This book was the result of a big push by one of my education mentors in life, Sandra Angles. She's the mentor with whom I bounce around ideas, get advice from, or just talk to become inspired about education. When she told me I needed to write a book about my art/math projects, I listened. I hesitated for a long time because I'm not a writer, and writing is my weakest skill (that's why I'm a math teacher)! Thanks, Sandy!

This project could not be successful without all the hard work by my students. They have met and exceeded my expectations in so many ways! Their imagination and creativity to take something so simple and make it "crazy cool" has, to this day, blown my mind! I look back at some of their work, and I am still amazed that they were the work of my students. Thank you, students!

Lastly, I want to thank Kristin McGregor for proofing and helping to edit this book into one coherent piece of work. There's no way this book could be published without her help! Thank you, Kristin!

About the Author

I graduated from UC San Diego with a bachelor's degree in Management Science and minors in Studio Art and Teaching. Afterwards, I returned to UCSD and earned a Master of Education and single subject teaching credential in mathematics. I also have a single subject credential in physics as well as supplemental credentials to teach art, economics, and business. For the last 14 years, I've taught at public schools, charter schools, private schools, and international schools. I've spent most of those years teaching at the high school level and a few at the middle school level.

My passion in education is in writing curriculum. I like creating curriculum that is meaningful, enjoyable, rigorous, and helps students retain knowledge long term. It's more difficult and time-consuming, yet the payoffs are amazing. I have won many grants, including ones from Toshiba, the National Council of Teachers of Mathematics, and the Friends of East County. These grants were written to win money to buy materials for projects which combined math with other subject areas such as science, art, and music.

During my spare time, I enjoy a quick hike, basketball, and social dancing. A little known fact about me is that I've taught swing dancing longer than I've taught math! In fact, it was teaching swing dancing that piqued my interest in teaching.

If you really want to know more about me, you are more than welcome to go to http://tanhuynh.oclassrooms.com to see my digital vitae.

Phew, I'm done writing about myself!

Introduction

One of the arguments against Project-Based Learning (PBL) by math teachers is the fact that we often see them as "fluff work." Most of the projects I've seen look amazing, and as a math teacher, I can totally see how the math is integrated. However, when I'm looking for the student's work in the creation of it, I find that it is minimal or sometimes non-existing. Furthermore, when asking the students about their projects, they often can articulate the application of the mathematics into the project; however, they cannot explain the actual mathematics that is relevant to the project. In other words, they were able to articulate the application of the concepts, but they failed to understand the "how" and "why" of the calculations.

What I find most difficult about PBL during the last 14 years of teaching is the balance between how much time the students spend on the math component and how much time they spend on the conceptual aspect of their project. When formulating ideas for a project in mathematics, I want to make sure that the math is the major component driving the project. The math is what creates the project versus a project that is sprinkled with math. The main goal I have in mind each time I create a new project is for students to practice enough of the math by doing as many problems as possible throughout the project.

The Gra$f$ It Art project came about as I was teaching one day and noticed a series of functions the algebra book used to graph a box. To the right is a similar image printed inside an algebra textbook. The students were asked to derive the four functions that are shown in red. Of course, this was an introduction to deriving linear functions. The chapters which preceded this example dealt with *graphing* linear functions. Students were now asked to look at functions that were already graphed and derive

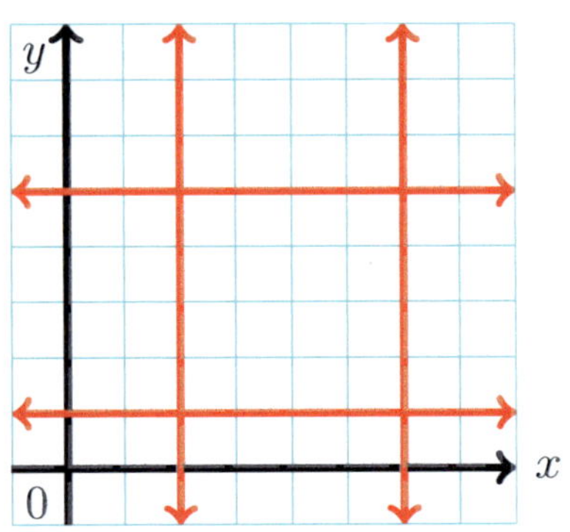

the actual functions.

What I saw in my mind was the following image to the right. I guess you can say that my art background often times kicks in to make boring things more interesting! By shortening all the lines and adding two more lines, the appeal of this problem increased ten-fold! This was no longer just a lesson on how to derive linear functions that graphed vertical and horizontal lines; it now became an art project that used math (specifically functions) as its medium.

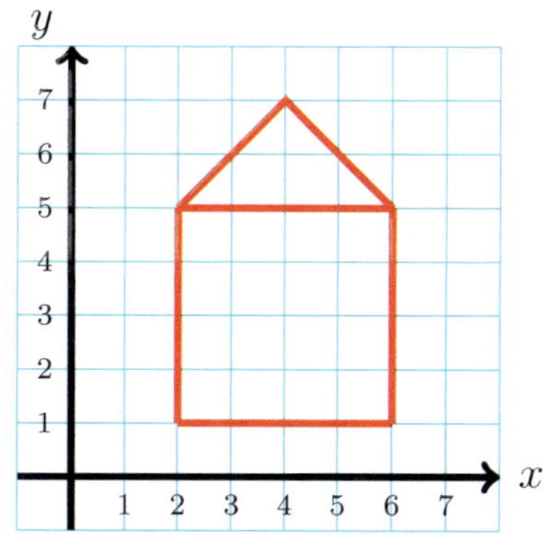

This started the creation of the Graph-It Art Project. I later learned that Graph-It Art is a company that sells embroidery artwork. I changed the project to be called Gra$f$ It Art. Of course, I had to be fancy about it and use a function notation in the middle of the word to symbolize the core nature of this project. It is limitless as to how far you can take this project. This project can be introduced as soon as pre-algebra and as late as calculus.

In case you're wondering what the functions are for the house above:

$$\begin{aligned} y &= 1 \quad \{2 \leq x \leq 6\} \\ y &= 5 \quad \{2 \leq x \leq 6\} \\ x &= 2 \quad \{1 \leq y \leq 5\} \\ x &= 6 \quad \{1 \leq y \leq 5\} \\ y &= x + 3 \quad \{2 \leq x \leq 4\} \\ y &= -x + 11 \quad \{4 \leq x \leq 6\} \end{aligned}$$

Simple, right?!

Before We Get Started

Teaching Styles

There are so many ways to approach teaching and every teacher has his or her own way of presenting and explaining his or her work. Since these are my projects, I will introduce to you the two main ways I approach teaching: direct instruction and pattern recognition.

Direct Instruction: Most teachers are familiar with this style of teaching. It's the most basic form of teaching, and it is the most direct way of teaching students concepts. I find that this method is effective in many cases. For instance, if I am to teach the formula for the area of a trapezoid, I will use a direct instruction style of teaching. I will present the formula and teach students how to use it. I use direct instruction when the focus of the lesson is on problem solving skills, rather than discovery. I would like them to learn the formula to apply to simple problems so that they can focus more on how to apply it to more complex problems. Consider the following problem. I would rather students figure out how to draw additional lines to decompose the diagram into four trapezoids and use the trapezoid area formula to compute the overall area than to spend time on discovering the formula itself.

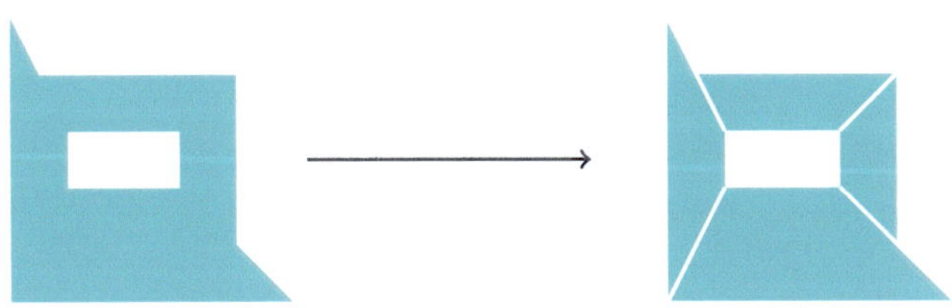

In my classes, approaches to solving problems are more important than deriving formulas. But that's not to say that there are times when I do want students to discover how a formula has been derived.

Pattern Recognition: I use the term "pattern recognition" instead of discovery-based approach because I find that the learning of mathematics has more to do with recognizing patterns than actually discovering it. I associate the word discover with expeditions and explorations. Call it what you may, it's my book, and I'm going to refer to this approach as "pattern recognition!"

This approach is quite simple. I tell the students at the beginning of the lesson, "I'm going to put up a series of problems of which I will work each out. Your goal is to copy each problem into your notebook and do the last 3 problems. From here on, I will not say a single word nor explain anything I am doing." Once I say that to the class, I solve 4-6 problems on the board, working out each problem thoroughly, boxing the answers for each problem. The last 2-3 problems are for the students to work out on their own. For example, if I want students to learn how to expand and solve summation problems, I would write the following 4 problems on the board, as well as 3 solvable problems.

1. $\sum_{i=1}^{5} i = 1 + 2 + 3 + 4 + 5 = \boxed{15}$
2. $\sum_{i=4}^{11} i = 4 + 5 + 6 + 7 + 8 + 9 + 10 + 11 = \boxed{60}$
3. $\sum_{i=1}^{3} 2^i = 2^1 + 2^2 + 2^3 = 2 + 4 + 8 = \boxed{14}$
4. $\sum_{i=1}^{5} 7 - i = (7-1) + (7-2) + (7-3) + (7-4) + (7-5)$

   $\sum_{i=1}^{5} 7 - i = 6 + 5 + 4 + 3 + 2 = \boxed{20}$
5. $\sum_{i=10}^{15} i =$
6. $\sum_{i=1}^{5} 5i =$
7. $\sum_{i=1}^{5} i(i-2) =$

This approach to teaching came about when I had a classroom full of English Language learners. I found that the more I talked, the more confusing it was for everyone. So, I decided to experiment with no talking at all. In fact, with the example I showed, I don't explain anything about the sigma notation. I let the students figure it all out. By the end of 10 minutes, most students will have it figured out. Those who are struggling are usually helped by students who figured it out (which they're very proud of doing). I usually end the lesson with a very brief direct instruction making sure that students understand certain terminology such as "sigma," "summation," "i-th term," etc. For me, this is one of my favorite ways of teaching, and I have been able to get through more material throughout the year as a result. In addition, students unanimously enjoy this way of learning. Students love puzzles, and pattern recognition is just like solving a puzzle–you're looking for the patterns.

Projects

I wouldn't call myself a pure PBL teacher. PBL takes up about 30% of what I do. I use it when it's appropriate. A large part of my teaching is still very teacher-centered. I still give daily homework, weekly quizzes/tests, midterms and finals. Students are expected to do most of a project at home. In addition, the project lags behind the curriculum that is currently being taught. Since the project is the application of the content being learned, it is *applied* when the students have understood how to do the problems out of context. The main goal of teaching for me is that the students understand how to do the math long term; projects are the icing on the cake.

When I have students work on projects, there are three things I focus on the most: 1) the students showing the math, 2) personalizing projects to the students, and 3) the public display of all projects.

Showing the Math: Make sure that a large component of the project consists of students showing the mathematical work leading up to the final product. This is the first thing I look for when I look at math projects. When this component is lacking, I consider the project a waste of time no matter how great it is presented. Furthermore, I'm not looking for a few mathematical equations here and there; I'm looking for at least 25 problems leading up to the final product, and for each problem, I'm also looking for the work leading to any final answers.

For instance, if a student wants to derive the function of the left side to a roof of a house (shown in red), I would want to see the following:

$$A(1,3) \text{ and } B(7,6)$$
$$m = \frac{\Delta y}{\Delta x} = \frac{6-3}{7-1} = \frac{3}{6} = \frac{1}{2}$$
$$y = \frac{1}{2}x + b$$
$$(1,3) : 3 = \frac{1}{2}(1) + b$$
$$\frac{5}{2} = b$$
$$y_{AB} = \frac{1}{2}x + \frac{5}{2}$$
$$D : [1,7], R : [3,6]$$

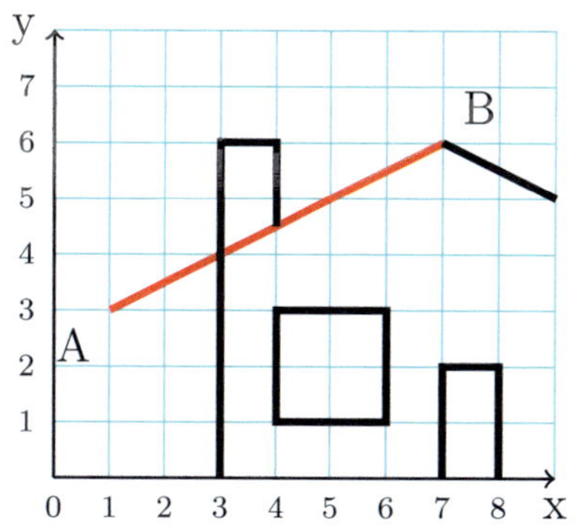

This would be one of many problems as part of the "work" section in an accompanying folder/binder. The work should be polished work demonstrating a professional look that is coherent and articulate.

Personalization: The projects presented in this book should merely serve as a soundboard for creating unique projects of your own. Take them and add or remove things that you like or don't like. Give them the unique flavor of your class. For instance, you may like the idea of a Skyline Project, but if your students don't live anywhere near city skylines, this may be difficult for your students to relate and find interest in. Rather, change the project to graph your school from different perspectives; the students are still practicing all the required skills, and they can now relate it to something more tangible. For example, have them graph the image of a landmark building within the town. Either way, make the project more personal to your students.

Community Projects: Every year I try to do a community project where each individual student's work contributes to an overall larger "community project." It is crucial that students complete their project because the overall display would be incomplete. In fact, students will hold each other accountable for completing this project because everyone is looking forward to how the larger project will look. Imagine displaying a completed jigsaw puzzle with one piece missing. And this is the analogy (as well as an example I show in class) I make when introducing this type of project. One important note here is that the teacher should also be a part of the community project. For instance, creating your own skyline that will fit in with all the other students. This is another reason why you should do the project before or concurrently with the students. "Community" is a central theme in all of my classes.

Public Gallery: One of my favorite parts of doing a project is the public display of the students' work. I try to find a local art gallery that is willing to exhibit student work for a week or even a few days. I make it a big deal, making fliers and invitations, calling parents, administrators, and everyone I can to come view the projects on opening night. On opening night, I try to have music and food; in addition, I convince the students to dress up for the event. Students who know that their project will be displayed for the public to see, and especially their parents and teachers, are 10 times more likely to do an amazing job. In fact, I have had students' work purchased for as much as $500 after the exhibition. Not bad for an eighth grader! Unanimously, students comment about how proud they are of their work when it gets displayed in a public gallery.

Tests and Quizzes At the end of each week, I still give a test or quiz on the content that was taught throughout the week. I understand that in many schools, "progressive" education centered around PBL discourages formal assessments such as tests and quizzes. I believe that formal tests and quizzes are extremely important tools to check for student understanding as well as helping students to develop the skills necessary for future formal assessments. Tests and quizzes count at least 60-70% of a student's overall grade. At the end of the day, my students need to understand and remember long-term the material they have learned.

How to Use this Book This book contains a complete project for students learning the coordinate system, usually taught in Pre-Algebra. The Basic Skyline project is divided into four different parts: The Design, The Functions, The Digital Design, and the Reflections. Each part explains what materials are needed, directions for the teacher, and lessons to be taught. Your pacing should be dependent on how quickly the students understand each lesson in order to produce a successful project. I usually spend about a month on the project so that students can really put forth their best work. However, that does not mean that I spend a month of in-class time to do the project. The amount of in-class time spent on the project will last as long as I can get through the lessons successfully. Most of the work takes place outside of class, in addition to any other regular assignments I may give them. Project work is meant to also teach students long-term project time management. Since this is something they're learning, make sure to help them develop this skill. You may tailor this project to fit whatever theme or creative avenue you would like to take. Most importantly, have fun with the students with this idea of using graphing as a art.

Let's Get Started

This book is the first in a series of short books where each project gets increasingly complex. Each is written for a specific level of mathematics. This book focuses on Pre-Algebra skills. The project is designed to be a stand-alone project or steps to building bigger and more complex projects. Use it how you want to, but keep in mind two things when doing projects with students: let the math drive the project and let the students drive the project. What keep students engaged is the fact that it is *theirs* and *they* want ownership of their own learning.

Things you'll need to do

- Get familiar with Desmos (http://www.desmos.com), an online graphing program. Learn how to add simple to complex equations, as well as constraints for functions. In addition, learn how to set graphing parameters for axis and points. Create your own account and have the students each create their own account.

- Do the project yourself. One of the things I find interesting about PBL is that almost all of the time, I've never seen a teacher's version of the project. It is *absolutely important* that the teacher does a version of the project beforehand or with the students. As a teacher, you will discover problems and difficulties that the students may have with the project. In addition, doing your own project allows the students to have a project to model after, both in quality and creativity. Make sure you try to create the best project possible as this sets the standard for your students. This will challenge the top students in class to outperform you. Always share your progress with your students (if you're doing the project alongside them). I have heard fewer complaints from students when I participate by doing my own version of the project.

- Determine how you will showcase your projects. Whether you set up a gallery for public viewing or present the projects via the TV announcements or at an assembly, make sure that the projects are showcased to as many people as possible. This culture of public display of student work is extremely important. When students know that their projects are up for public viewing by other students, teachers, administrators, or to the public, they will work harder to create the best project.

- Relax and don't stress out! PBL is supposed to be fun. If you're having a fun time creating a project of your own and discovering cool ways to do it, the students will catch on to this enthusiasm and will love the project too! It's that simple! Make the class into a community of "arti-maticians" and the students will forget they're even doing math!

Project 1 - The Basic Skyline

$Level : Pre - Algebra$

$Time : 1\ week$

$Lessons : 4$

Prerequisites:

- Familiarize yourself with Desmos (www.desmos.com)
- Make sure students have a good grasp of graphing horizontal and vertical lines. In particular, students should understand that:
  - all horizontal and vertical lines have the general equation, $y = n$ and $x = n$, respectively
- Students should be familiar with the domain and range of horizontal and vertical linear functions.

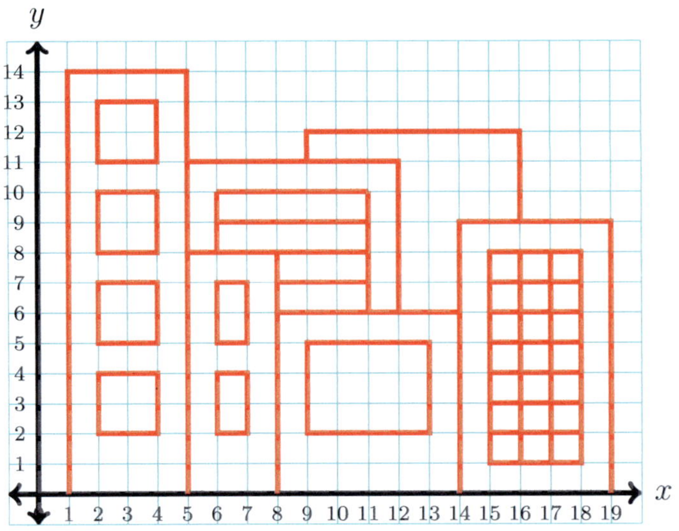

Part I - The Design (Draft 1)

I usually introduce the project even before I teach the first lesson pertaining to this project. This first part requires no knowledge of graphing. It's a fun assignment for students to do to pique their interest about the project.

What's Needed:

- **Graph Paper:** I recommend handing out the same piece of graph paper to all the students. When you display the projects later, they'll look more uniform and appealing. Decide which quadrant(s) you want the students to make their design. Generally, I use the first, or the first and second quadrants. It's best to "design" this graph paper for the students.
- **Ruler:** I mark down students for not using a ruler. Drawing without it is sloppy and is never acceptable in my class.
- **Pencil:** Using a pencil allows students to make corrections more easily.

Directions:

- **You** should show several examples of work done by other students and ones you've done. You should set certain requirements such as:
  - The number of buildings (in front and behind)
  - Having windows on the buildings
  - Something extra, such as a sign or billboard

  If you don't do this crucial step, you may have a few students constructing only one or two buildings because they didn't know what the expectations were. In addition, they may not even know what a skyline is. Showing examples sets the expectations for the students. You may show any of the examples from my website (grafitart.oclassrooms.com).
- **The students** should design a skyline using just horizontal and vertical lines.
- **Post** an example on your course website or in your class as to what a final product would look like. Show this to the students and go over the expectations you have for the project. Let the students understand how they can earn full credit for the assignment.

Lesson 1

Functions of Horizontal and Vertical Lines

Instructions

I like to teach these simple lessons by having students recognize patterns rather than direct instruction. In this case, I would have the diagram and lines already drawn by the time the students are settled into class. Then, I would proceed to write the equations of the horizontal line for 4 of the 7 lines. I would leave the last three for students to do.

Horizontal Lines

Examples to write on the board (without verbal explanations):

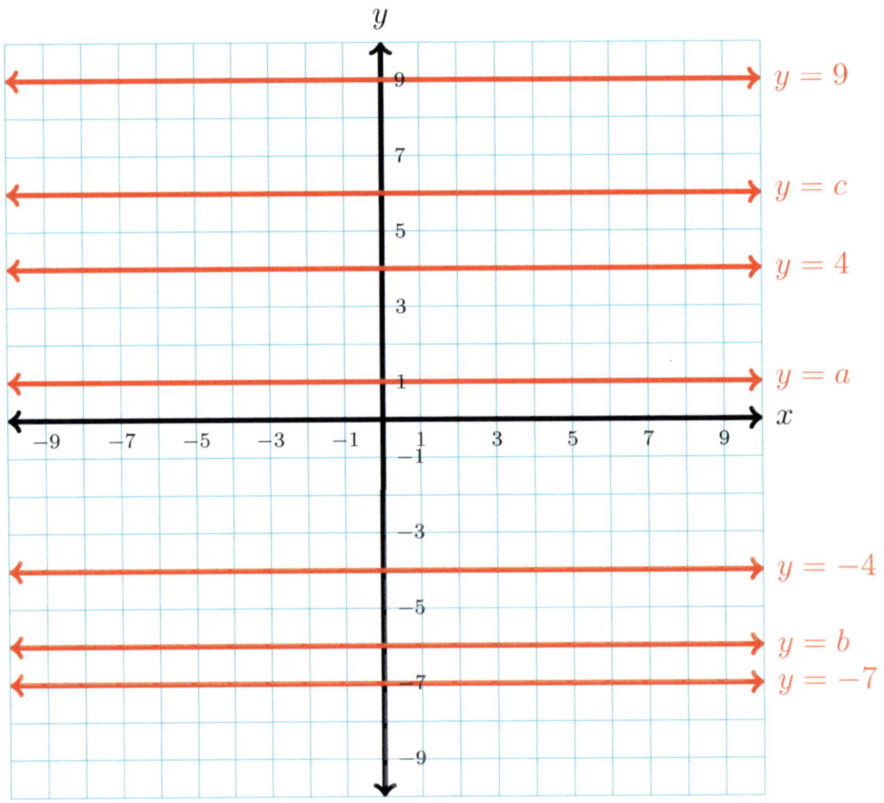

Have students determine the values of a, b and c. Almost every student will do this correctly.

Vertical Lines

Examples to write on the board (without verbal explanations):

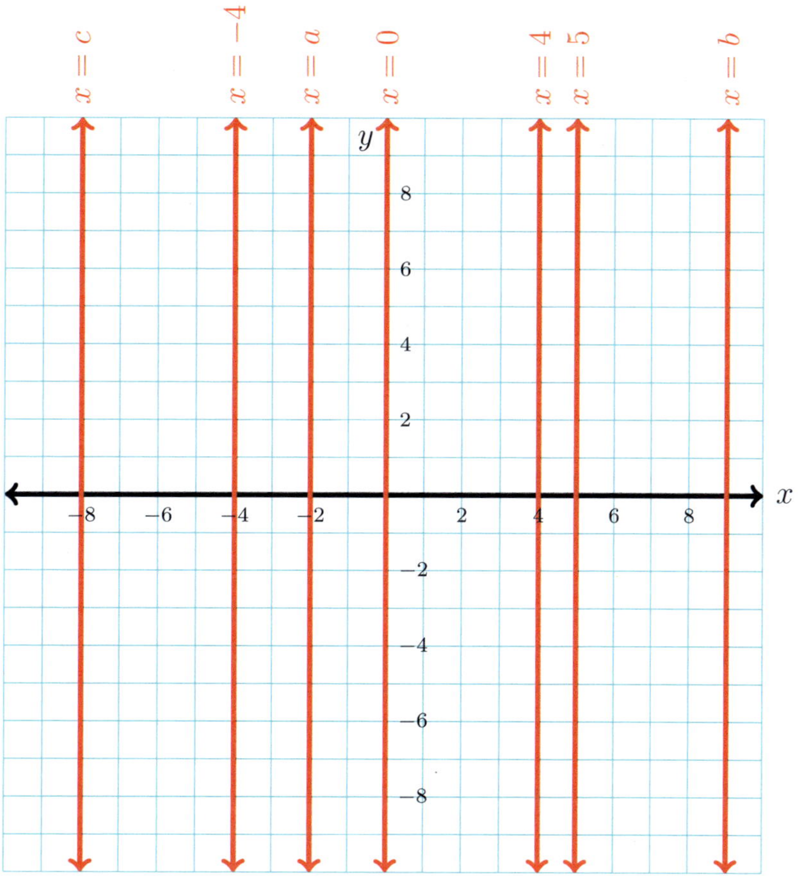

Have students determine the values of a, b and c. Almost every student will do this correctly.

Part I - The Design (Draft 2)

I believe in multiple drafts when it comes to doing projects. Much like an English paper, the first draft is nothing close to polished work. Students should understand that this is part of how companies approach projects and designs. After collecting the students' first draft, you should have students determine what they need to do to improve their second draft.

What's Needed:

- **Graph Paper:** Give students a second copy of the graph paper you handed out the first day. Drafts should have similar templates. Don't let students add on to the first draft; you want to see progress.
- **Ruler:** Again, drawing straight lines without the use of a ruler is sloppy and unprofessional.
- **Pencil:** "Drafting" is usually done with a pencil.

Directions:

- **Show** examples again to the class. Ask students how they can improve their design to make it more creative, unique, and to meet the expectations you have set. Show the class examples that meet or exceed your expectations.
- **Point out** what specific improvements would make their design better. Here are some helpful ideas:
  - Overlapping buildings to give the skyline a more natural look
  - Windows on buildings, instead of blank facades
  - Billboards and other small details

***Jigsaw Project** The Basic Skyline Project is a perfect project to use as a *jigsaw project.* If you want to extend this project into one, this is the time to do it. Give students a random number from 1 to the total number of students in the class (plus yourself). Once students have their number, they must work with a student one number less than them and one number greater than them. For instance, if you picked 9, you will have to work with the student who has the number 8 and the student who has the number 10. On the west side of your skyline, you must create a building that connects to a building on the east side of student number 8. In addition, you must create a building on the east side of your skyline that will connect with a building on student number 10's skyline.

This jigsaw is great because in the end, you'll create a skyline that is connected from student number 1 to the last student in class. The students will name their city and you will see a string of city skylines all connected. This is why it is important that all the students use the same style of graph paper. This type of project reinforces the central theme of *community* in my class.

See the example below:

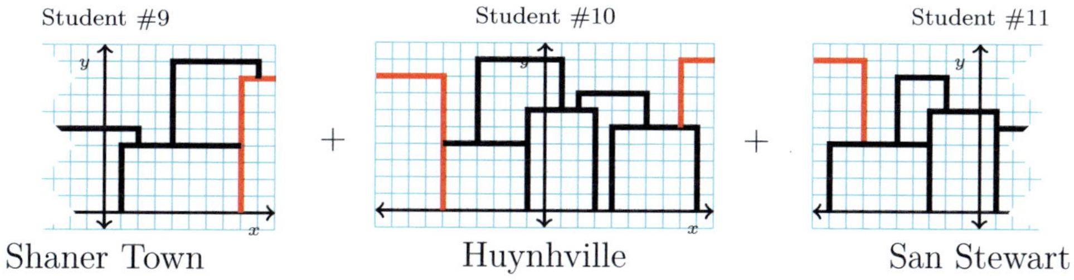

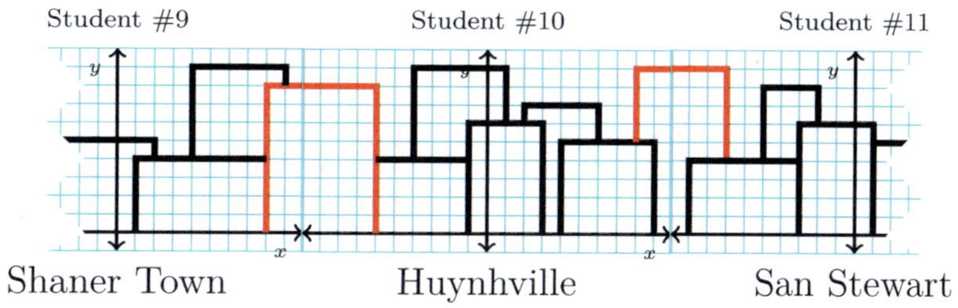

The important part of a jigsaw project is to promote classroom community and learning.

Lesson 2

Domain and Range of Horizontal and Vertical Lines

Instructions

Like the previous lesson, I like to use the pattern recognition approach. However, notice that this time I included along the x- and y-axis, $-\infty$ and ∞. This is important because you want students to make the connection of the axis continuing on to infinity in both directions.

Horizontal Lines

Examples to write on the board (without verbal explanations):

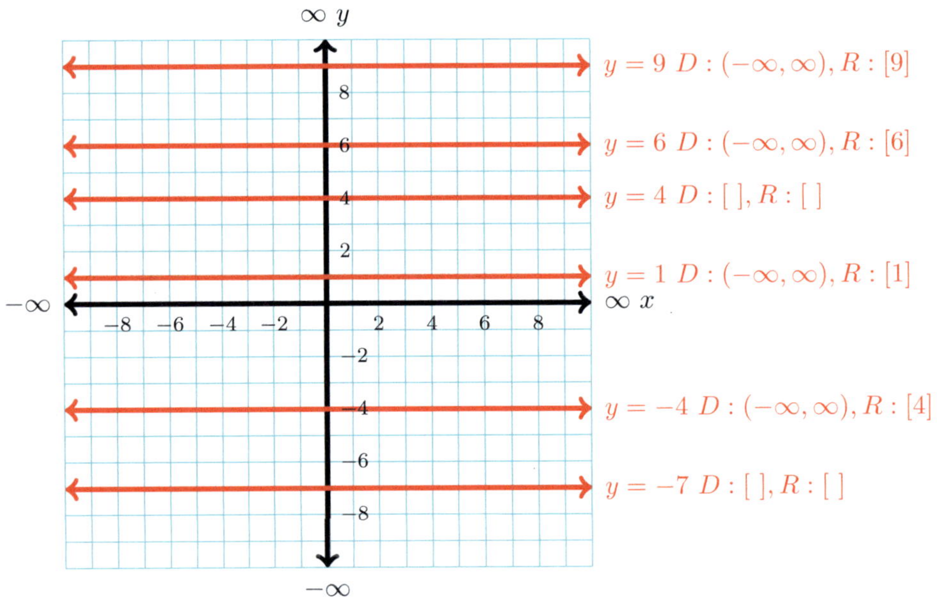

Have the students complete the missing domain and range. A short discussion should follow regarding the following:

- Why do all the horizontal lines have the same domain?
- What is the difference between using brackets, [ ], versus parenthesis, ()?
- Discuss with the class the proper use of interval notations.

Vertical Lines

Examples to write on the board (without verbal explanations):

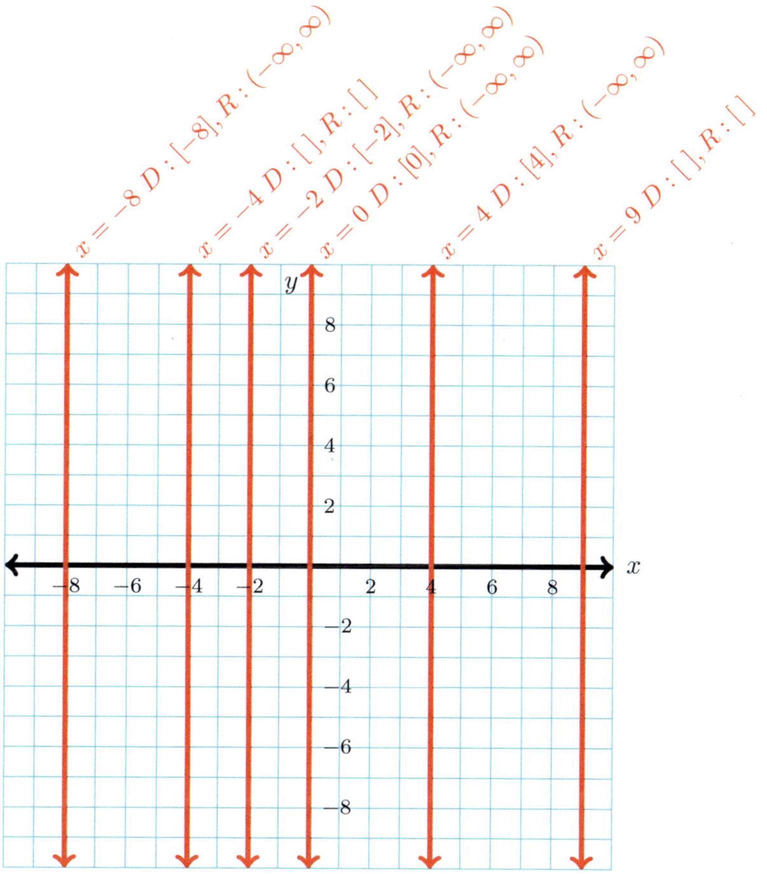

Have the students complete the missing domain and range. Again, a discussion should follow this regarding interval notation, including the similarities and differences between the domain and range of horizontal and vertical lines.

Part I - The Design (Draft 3)

The third draft should be the most polished since this will be the draft that the students will use to derive all the linear functions. It is important to let students know that they will have to do a lot of work if they make major changes later on. If you're using this as a jigsaw project, make sure the students draw the connecting buildings correctly and that everything lines up correctly. Common mistakes usually occur when the connecting building is a "front" building on one student's skyline and a "behind" building of another student. Make sure to point this out to students. What's needed:

- **Graph Paper:** Again, use the same style graph paper since day one.
- **Ruler:** Hopefully, the students understand the importance of using a ruler by now.
- **Pencil:** Keep them sharpened! They'll make better lines!

Directions:

- **Show** several examples of work done by students in class. Let a few shine and help to inspire others to make their projects even better. A little show and tell is definitely useful at this stage.
- **Teach** students how to critique each other's work in constructive ways. Show them proper ways of encouraging other students to work harder.
- **Encourage** students to work together and help each other create the best panoramic skyline. Make it a healthy competition between classes to see who will make the better skyline.

Geometry Extension I've seen in the past a couple of teachers use this moment to create a mini lesson on the area of rectangles and irregular shapes. They had the students find the area of the facade of the buildings. I've even seen teachers asking students what percent of a building is showing to the viewer. These are quite impressive extensions to the project. Again, give this project a twist and find teachable moments!

Lesson 3

Domain and Range of Horizontal and Vertical Lines, Rays and Line Segments

---

Instructions

Like the previous lesson, I like to use the pattern recognition approach. However, notice that this time I included along the x- and y-axis, $-\infty$ and ∞. This is important because you want students to make the connection of the axis continuing on to infinity in both directions.

Horizontal Lines

Examples to write on the board (without verbal explanations):

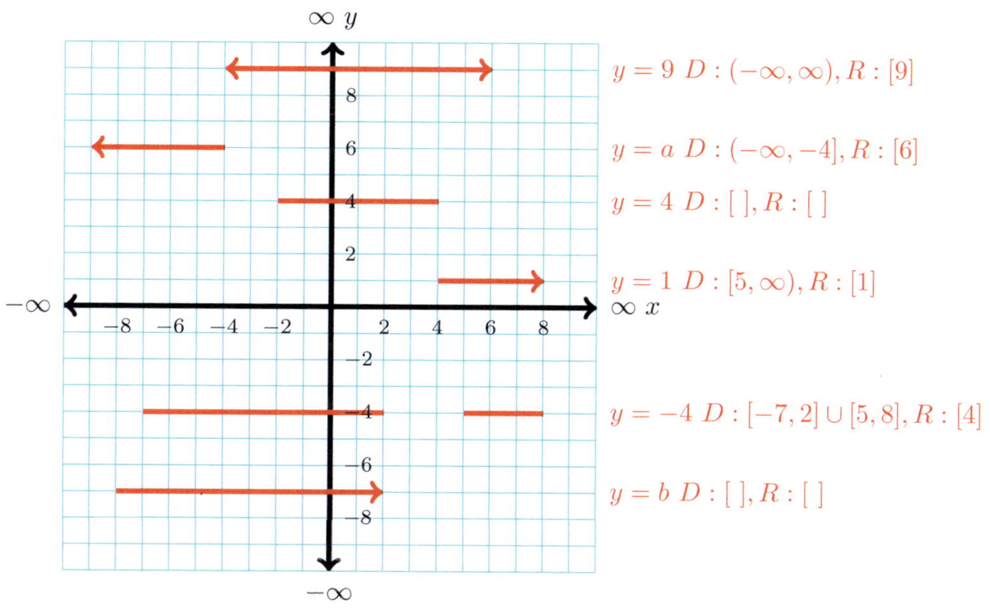

Have the students find the value of a and b as well as complete the missing domain and range. A short discussion should follow regarding the following:

- Why do we use/do not use [or] for infinity?
- Are there any cases when (or) is used for a number that is not infinity?
- What is the symbol $\cup$ and when is it used?

Vertical Lines

Examples to write on the board (without verbal explanations):

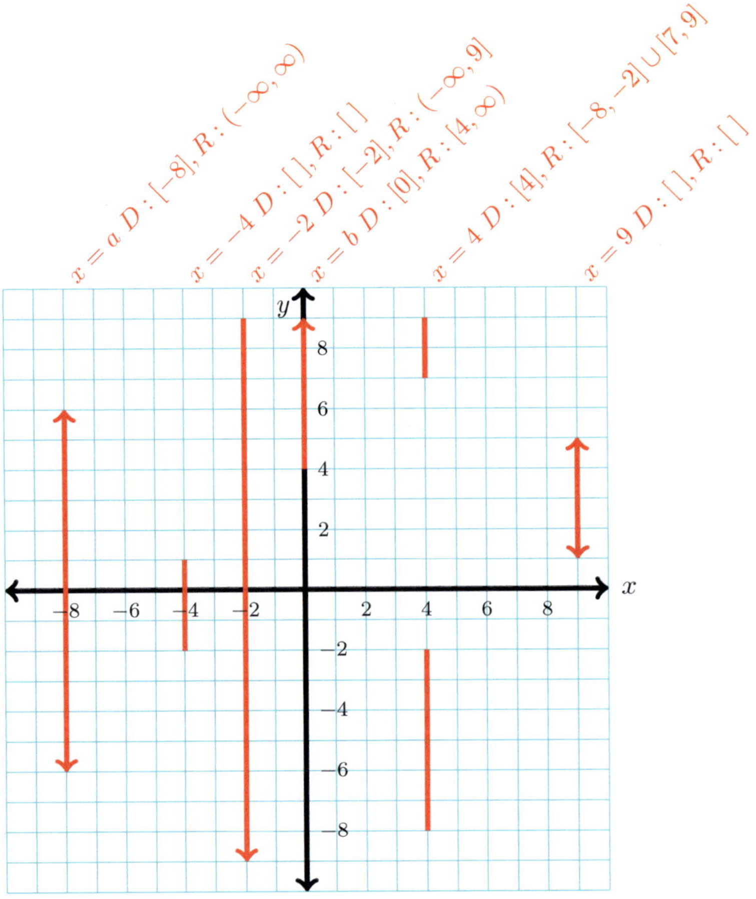

Have the students find the value of a and b as well as complete the missing domain and range. Again, a discussion should follow this regarding interval notation, including the similarities and differences between horizontal and vertical lines.

Part II - The Functions

This is my favorite part of the project where the content being taught applies directly to the project. The students will now take their design and find the equation, domain, and range of each line segment. However, you want to make sure everything is organized from here forward or else you'll have a frustrating time helping students.
What's Needed:

- **Draft 3:** By now, this draft should be quite polished. Have the students label all endpoints of all line segments in their design. Start with A-Z and then use "AA, AB, AC,...,AZ." From there use "BA, BB, BC,...,BZ." Continue this notation until all endpoints are accounted for. See the example:

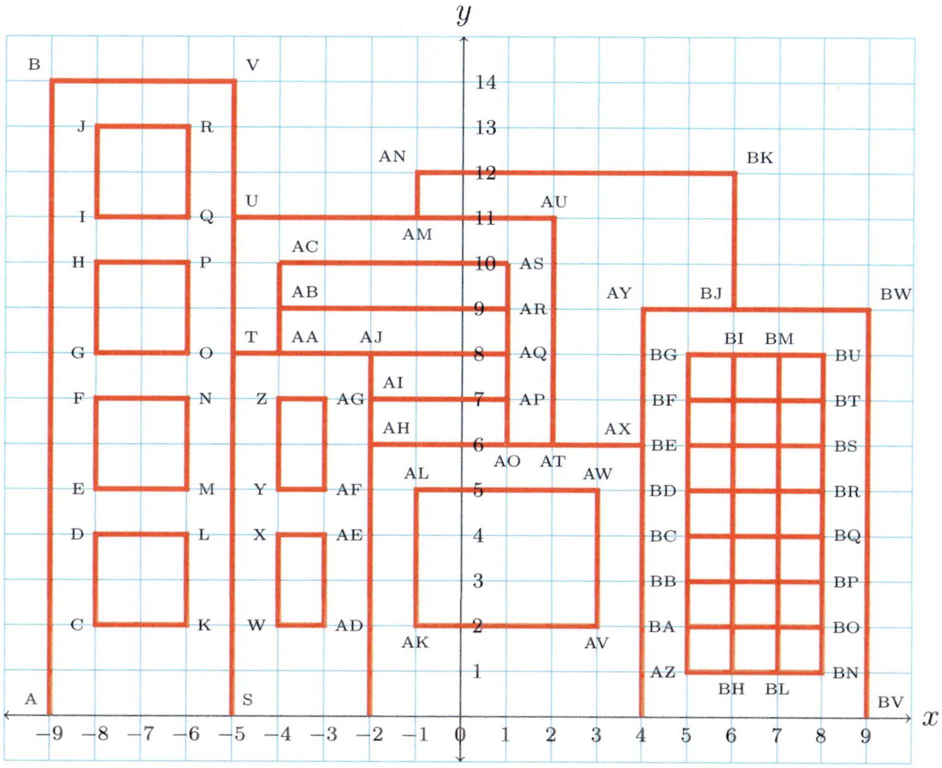

  Notice in the image that the cross hairs of the building on the right are not labeled. That's because these are not endpoints to a line segment.

- **Computer:** Having students handwrite this section is not a good idea. Teach the students how to use Google Doc, MS Word, or any other Software to make tables and write equations.

- **List of Coordinates** I personally like using MS Word since you can create nice tables for lists. You should have a maximum of 4 columns to a page. I make sure the student titles the page "Skyline Coordinates." Coordinates should be listed in alphabetical order. For instance:

Skyline Coordinates			
A (-9,0)	AA (-4,8)	BA (5,2)	
B (-9,14)	AB (-4,9)	BB (5,3)	
C (-8,2)	AC (-4,10)	BC (5,4)	
...	...	...	...

- **List of Functions** Again, I like using MS Word for this since it creates nice tables, and there's a built in equation editor (I recommend spending a little time teaching the students how to use the equation editor in MS Word). Make sure the students title this page, "Skyline Functions." I like having students separate the functions into two lists, horizontal and vertical. This will make things easier when they have to type all of it into the graphing software. I have the students create a table like the following:

Skyline Functions - Vertical Lines			
Line Segments	Functions	Domains	Ranges
A-B	$x = -9$	{x=-9}	{0 ≤ y ≤ 14}
C-J	$x = -8$	{x=-8}	{2 ≤ y ≤ 4, 5 ≤ y ≤ 7, 8 ≤ y ≤ 10, 9 ≤ y ≤ 11}
K-R	$x = -6$	{x=-8}	{2 ≤ y ≤ 4, 5 ≤ y ≤ 7, 8 ≤ y ≤ 10, 9 ≤ y ≤ 11}
S-V	$x = -5$	{x=-9}	{0 ≤ y ≤ 14}
W-AC	$x = -5$	{x=-4}	{2 ≤ y ≤ 4, 5 ≤ y ≤ 7, 8 ≤ y ≤ 10}
...	...	...	...

Notice that the function for C-J is a vertical function with multiple ranges. Students do not have to write four different functions between point C and point J. If students understood the preceding lesson, then this shouldn't be a problem. This saves a lot of time and effort. In addition, this is a great way for students to be introduced early to piecewise functions. Also, notice that the domains and ranges are not written in interval notation. That's because Desmos (the website I use for this project) uses this style of notation for domains and ranges.

Directions:

- **Show** one really good example of how both the list of coordinates and list of functions should look. If this is your first time doing this project, show one that you have done. Set the standard for how professional you want your students' work to look.

- **Teach** students how to use a word processor like MS Word or Google Doc to:

  - Use the header to include their name, date, class, etc.
  - Create a table with multiple columns.
  - Insert equations, especially the use of subscripts and symbols like $\cup$.
  - Format the table to make it professional-looking in the end.

- **Emphasize** that organizing this section correctly will help them with the next part of the project.

Grading Homework Don't worry about knowing whether the students got every function correct. I usually spot-check their homework for completeness. They'll have to input these functions into Desmos.com later. If the image shows up in Desmos exactly like Draft 3, then they did it right! No need to spend hours grading. Don't worry, I have it all figured out for you!

Project Binder I often ask students to buy their own binder with dividers for each major project. For this project, there are just four sections: 1) The Design Drafts, 2) The Functions, 3) The Digital Design, and 4) The Reflections. The students should have three drafts for the first section. In section 2, the students will have several pages listing all their coordinates and several pages of functions with their given constraints. Section 3 will include an image, a link, and a QR code for their work completed on www.desmos.com. The last section will be discussed later.

Part III - The Digital Design

When the submissions start coming in, you're going to be proud of what the students have accomplished. Some students will take it to a level totally unexpected, using over 400 functions to graph their skyline. These will become the ones you'll put up in your class forever.

What's Needed:

- **Draft 3:** The students will use this to check to see if the line segments in Desmos is matching up with their own design. If a student made a mistake on their list, they will work hard to fix it (most of the time without you telling them to).
- **List of Functions:** Make sure students correct any work they may have done wrong after checking with Desmos. In addition, if students want to make additions to their design, have them update the list of coordinates and list of functions. The number of equations on Desmos should match the number of equations on their typewritten list.
- **Desmos:** Remember to familiarize yourself with the software before you start teaching it to the students. Play around with it and do a mini or full project before starting with the students.

Directions:

- **Show** students how to input equations into Desmos as well the corresponding domain or range. Sometimes mistakes can be as simple as not typing a comma or equal sign. Students can get frustrated quickly. Help them learn how to troubleshoot their own work.
- **Teach** students how to send their work to you via the share link in Desmos.
- **Post** a few examples of student work from Desmos. If this is your first time, use some of the ones from my website. Again, setting high expectations will allow students to work harder than they think they can.

Lesson 4

Using Desmos.com as a Grapher

Instructions

Unlike previous lessons, I use a lot of direct instructions when teaching students how to use new software. Hopefully, you have already familiarized yourself with Desmos.com.

Directions

1. Make sure each student registers for an account on Desmos.com. This is important because they'll need an account to save their work and send it to you. Have students write down their username and password in a safe place.

2. Demonstrate how to set the $x-$ and y-axis. While you're doing this, also place a check-mark into the "Square Grid" option. The software may adjust the domain and range a little, but your image will come out looking better.

3. Instead of having students input equations from their lists immediately, have them go through a simple example with you first. On the next page, I will run through an example I do in class with the students. There are a few things students should understand when entering equations into Desmos. Students will only need to specify the domain when they're graphing a horizontal segment (and only the range for vertical line segments). In addition, the way that Desmos recognizes domains and ranges is in piecewise notation $\{a \leq x \leq b\}$.

4. When the students are done, have them share their graph by clicking the icon in the top right hand corner and entering your school email address. I also have them turn in their hand-graphed (Draft 3) as well. I grade the students based on whether their Desmos graph looks exactly like their hand-drawn Draft 3.

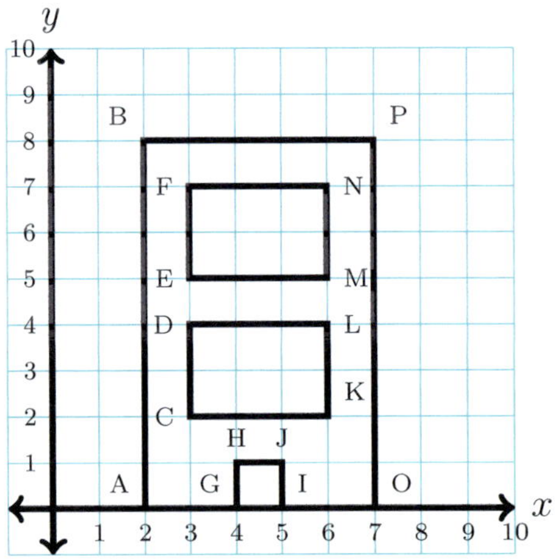

Using the above example, let's run through a few entries:
Line segment A to B would be entered into Desmos as:

$$x = 2\ \{0 \leq y \leq 8\}$$

Line segment C to F would be entered as:

$$x = 3\ \{2 \leq y \leq 4, 5 \leq y \leq 7\}$$

Notice the multiple entries for range. If the students created the list of functions correctly, they won't have to repeatedly input the same function with different domains/ranges. The rest of the functions can be entered as follows:

$$x = 4\ \{0 \leq y \leq 1\}$$
$$x = 5\ \{0 \leq y \leq 1\}$$
$$x = 6\ \{2 \leq y \leq 4, 5 \leq y \leq 7\}$$
$$x = 7\ \{0 \leq y \leq 8\}$$

And for the horizontal line segments:

$$y = 8\ \{2 \leq y \leq 7\}$$
$$y = 5\ \{3 \leq y \leq 6\}$$
$$y = 2\ \{3 \leq y \leq 6\}$$
$$y = 7\ \{3 \leq y \leq 6\}$$
$$y = 4\ \{3 \leq y \leq 6\}$$
$$y = 1\ \{4 \leq y \leq 5\}$$

If everything was entered correctly, you should have an image on Desmos that looks like the image above (without the labeled points).

If you followed the example, your Desmo.com graph and list of equations should look very similar to the following screenshot:

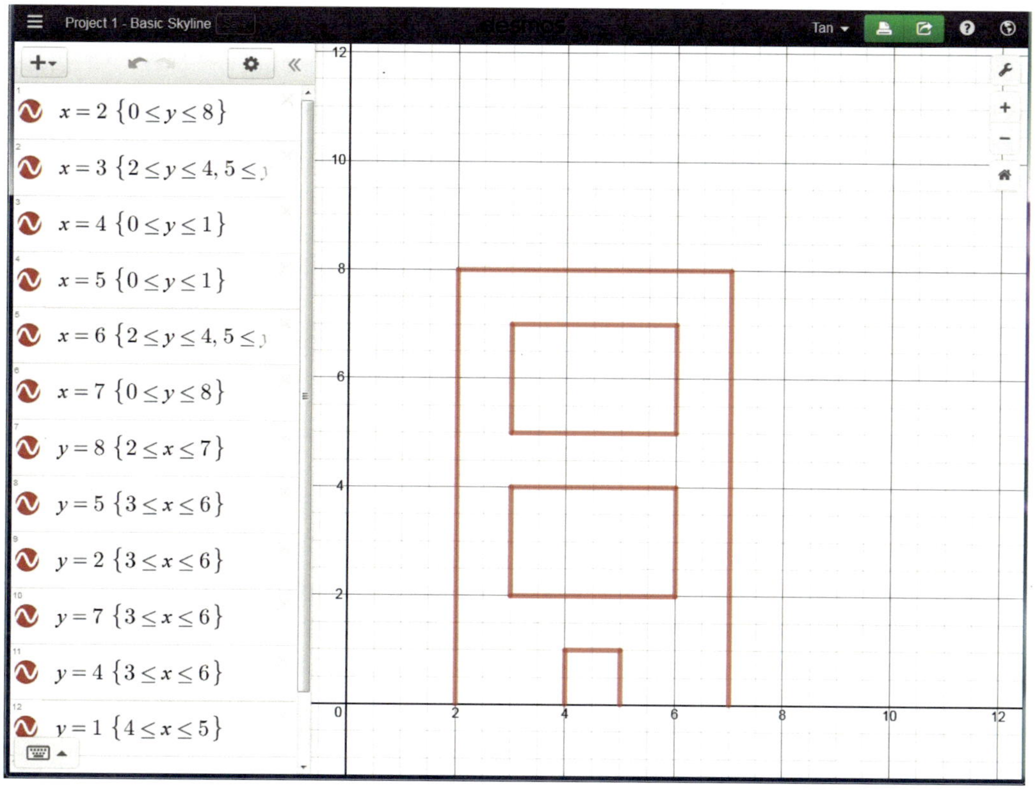

You should pay attention to two important icons. The first is the "cog" above the list of equations. This is where you can change your lines to be a different color or style of the line. The second icon is the "wrench" at the top right hand corner. This allows you to change the viewable area and other properties of the coordinate plane.

Part IV - The Reflections

I always have a reflection piece with all of my projects. Here, I get to ask students about their thoughts and feelings about the project. I ask questions about what they like and what they didn't like about the project. Students are usually very candid and will let you know their true feelings. The length of their reflection is entirely up to you. Some reflection questions I use are:

1. What did you like and dislike about this project?
2. What was the most frustrating part(s) about this project?
3. What would you recommend to change to make this project better for future students?
4. What would you do differently to make your project better?

What's Needed:

- **Parts I, II, and III:** Students should have these parts already in a binder/folder with dividers. As they look back at what they have done, it'll make it easier for them to realize what they've learned, what they like, and what they didn't like.
- **Computer:** Not much in my class is handwritten when it comes to projects. Make sure students have access to computers to complete this part.

Directions:

- **Show** students examples of good responses versus bad responses. Give students sentence starters such as "I like...about the project because..." or "I would have done...differently because...." Constantly model for students what appropriate writing looks like.
- **Display** sample projects so that students know how to organize their own. If this is your first time, create one of your own and use that to set the standards and expectations for the class.
- **Grading** is entirely up to you. I grade each part based on whether students followed directions and how complete the project is. However, I weigh the math part as 80% of the project grade. Make sure to let students know how you're grading upfront; they need to know what they will lose points for.

Appendix

Graph Paper I recommend teachers provide students the same graph paper for their designs for every draft of the project. You will end up with a mess if you have students purchase their own graph paper. If you're going to showcase the projects later, uniformity will be more aesthetically pleasing. You can find a graph paper template I used with my students on my webpage (grafitart.oclassrooms.com).

Tables I recommend that everything is done on computer using a word processor with tables. The following pages can give you an idea of a setup for students. My recommendation is to set up a template and allow students to download it. You can find a table template I used with my students on my webpage (grafitart.oclassrooms.com).

Samples Please have a project that exemplifies excellence. I always believe in the adage, "students will rise to the level of expectations." Having a model project is very important because students will see how well it's put together, and they will emulate how an excellent project should be organized. I always recommend the teacher to do one of his or her own and use that as the model. You can also find a sample project on my website (grafitart.oclassrooms.com).

Repeat directions and model for the students often. Students learn best when they hear things over and over, especially things you want them to remember. It is also why I repeat myself again and again throughout this book; I want to make sure you understand and are able to execute this project with ease. I want you to have fun and have a successful project!

Made in United States
Troutdale, OR
11/03/2026

56844701R00024